# 1

## one

# 2
## Two

# 3

# Three

# 4

# Four

# 5

# Five

# 6

## Six

# 7

# Seven

# 8

# Eight

# 9

# Nine

# 10
# Ten

How many Bananas can you see?

What colour is the Banana?

# How many Apples can you see?

What colour are the apples?

# How many Cherries can you see?

What colour are the cherries?

# How many Raspberries can you see?

What colour are the raspberries?

# How many Pears can you see?

What colour are the pears?

# How many Blueberries can you see?

What colour are the Blueberries?

# How many Oranges can you see?

What colour are the oranges?

# How many Peaches can you see?

What colour are the peaches?

# How many Strawberries can you see?

What colour are the Strawberries?

# How many Grapes can you see?

What colour are the grapes?

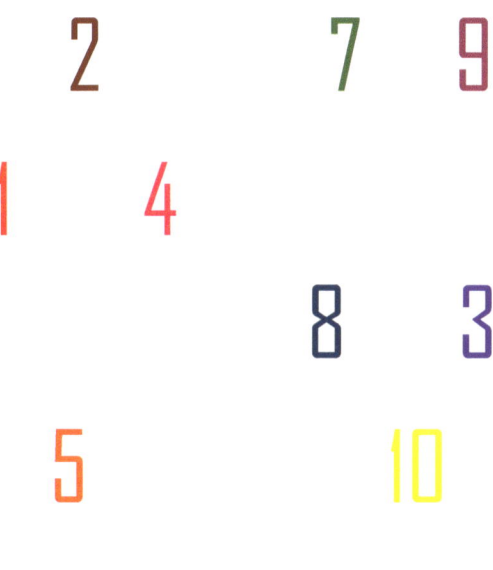

Can you point to number 1?

Can you point to number 2?

Can you point to number 3?

Can you point to number 4?

Can you point to number 5?

Can you point to number 7?

Can you point to number 8?

Can you point to number 9?

Can you point to number 10?

What shapes can you see?

What colours can you see?

Are there any objects in your home that are the same shape?

Shapes are all around us try looking for shapes that are the same in your home.

What Shape is the window in your home?

Can you spot the same shape in one of these pictures?

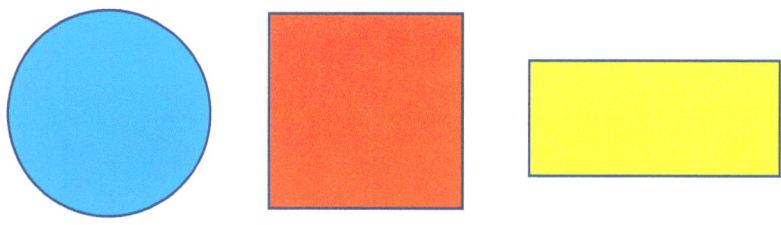

What shape is the fridge in your home?

Can you spot the same shape in any of these pictures?

What shape is the table in your home?

Can you spot the same shape in one of these pictures?

What shape are the food bowl in your home?

Can you spot the same shape in one of these pictures?

www.ingramcontent.com/pod-product-compliance
Lightning Source LLC
Chambersburg PA
CBHW041945240526
45473CB00033B/614